EMERGENCY PREPAREDNESS

EMERGENCY PREPAREDNESS

PAUL V. CREIGHTON
EDITOR

Nova Science Publishers, Inc.
New York

For permission to use material from this book please contact us:
Telephone 631-231-7269; Fax 631-231-8175
Web Site: http://www.novapublishers.com

LIBRARY OF CONGRESS CATALOGING-IN-PUBLICATION DATA

Available upon request

ISBN: 978-1-60692-510-2

Published by Nova Science Publishers, Inc. ✝ New York

CONTENTS

PREFACE

Potential terrorist attacks and the possibility of naturally occurring disease outbreaks have raised concerns about the "surge capacity" of the nation's health care systems to respond to mass casualty events. GAO identified four key components of preparing for medical surge: (1) increasing hospital capacity, (2) identifying alternate care sites, (3) registering medical volunteers, and (4) planning for altering established standards of care. The Department of Health and Human Services (HHS) is the primary agency for hospital preparedness, including medical surge.This is an edited and excerpted edition.

What United States Government Accountability Office Found*

Following a mass casualty event that could involve thousands, or even tens of thousands, of injured or ill victims, health care systems would need the ability to "surge," that is, to adequately care for a large number of patients or patients with unusual medical needs. The federal government has provided funding, guidance, and other assistance to help states prepare for medical surge in a mass casualty event. From fiscal years 2002 to 2007, the federal government awarded the states about $2.2 billion through the Office of the Assistant Secretary for Preparedness and Response's Hospital Preparedness Program to support activities to meet their preparedness priorities and goals, including medical surge. Further, the federal government provided guidance for states to use when preparing for medical surge, including *Reopening Shuttered Hospitals to Expand Surge Capacity*, which contains a checklist that states can use to identify entities that could provide more resources during a medical surge.

Based on a review of state emergency preparedness documents and interviews with 20 state emergency preparedness officials, GAO found that many states had made efforts related to three of the key components of medical surge, but fewer have implemented the fourth. More than half of the 50 states had met or were close to meeting the criteria for the five medicalsurge-related sentinel indicators for hospital capacity reported in the Hospital Preparedness Program's 2006 midyear progress reports. For example,

37 states reported that they could add 500 beds per million population within 24 hours of a mass casualty event. In a 20-state review, GAO found that

- all 20 were developing bed reporting systems and most were

* Excerpted from GAO Report GAO-08-668, dated June 2008.

coordinating with military and veterans hospitals to expand hospital capacity,

- 18 were selecting various facilities for alternate care sites,
- 15 had begun electronic registering of medical volunteers, and
- fewer of the states—7 of the 20—were planning for altered standards of medical care to be used in response to a mass casualty event.

State officials in GAO's 20-state review reported that they faced challenges relating to all four key components in preparing for medical surge. For example, some states reported concerns related to maintaining adequate staffing levels to increase hospital capacity, and some reported concerns about reimbursement for medical services provided at alternate care sites. According to some state officials, volunteers were concerned that if state registries became part of a national database they might be required to provide services outside their own state. Some states reported that they had not begun work on or completed altered standards of care guidelines due to the difficulty of addressing the medical, ethical, and legal issues involved in making life-or-death decisions about which patients would get access to scarce resources. While most of the states that had adopted or were drafting altered standards of care guidelines reported using federal guidance as they developed these guidelines, some states also reported that they needed additional assistance.

Why GAO Did this Study

Potential terrorist attacks and the possibility of naturally occurring disease outbreaks have raised concerns about the "surge capacity" of the nation's health care systems to respond to mass casualty events. GAO identified four key components of preparing for medical surge: (1) increasing hospital capacity, (2) identifying alternate care sites, (3) registering medical volunteers, and (4) planning for altering established standards of care. The Department of Health and Human Services (HHS) is the primary agency for hospital preparedness, including medical surge. GAO was asked to examine (1) what assistance the federal government has provided to help states prepare for medical surge, (2) what states have done to prepare for medical surge, and (3) concerns states have identified related to medical surge. GAO reviewed documents from the 50 states and federal agencies. GAO also interviewed officials from a judgmental sample of 20 states and from federal agencies, as well as emergency preparedness experts.

WHAT GAO RECOMMENDS

GAO recommends that the Secretary of HHS ensure that the department serve as a clearinghouse for sharing among the states altered standards of care guidelines developed by individual states or medical experts. HHS was silent on GAO's recommendation. HHS and the departments of Homeland Security, Defense, and Veterans Affairs concurred with GAO's findings.

ABBREVIATIONS

AHRQ	Agency for Healthcare Research and Quality
ASPR	Office of the Assistant Secretary for Preparedness and Response
CDC	Centers for Disease Control and Prevention
CMS	Centers for Medicare & Medicaid Services
DHS	Department of Homeland Security
DMAT	Disaster Medical Assistance Team
DOD	Department of Defense
EMTALA	Emergency Medical Treatment and Labor Act
ESAR-VHP	Emergency System for Advance Registration of Volunteer Health Professionals
HAvBED	Hospital Available Beds for Emergencies and Disasters
HHS	Department of Health and Human Services
HIPAA	Health Insurance Portability and Accountability Act of 1996
HRSA	Health Resources and Services Administration
HSPD	Homeland Security Presidential Directive
MRC	Medical Reserve Corps
PAHPA	Pandemic and All-Hazards Preparedness Act
VA	Department of Veterans Affairs

June 13, 2008

Congressional Requesters

The September 11, 2001, terrorist attacks on the World Trade Center and the Pentagon, the anthrax incidents during the fall of 2001, and the possibility of a naturally occurring disease outbreak or some other large- scale public health emergency have raised public awareness and concern about the ability of the nation's health care systems[1] to respond to bioterrorism[2] and other mass casualty events.[3] In a mass casualty event the ability of local or regional health care systems to deliver services consistent with established standards of care[4] could be compromised, at least in the short term, because the volume of patients would far exceed the available hospital beds, medical personnel, pharmaceuticals, equipment, and supplies.

Following a mass casualty event, health care systems would need the ability to "surge," that is, to adequately care for a large number of patients or patients with unusual or highly specialized medical needs. Providing such care would require the allocation of scarce resources and could occur outside of hospitals and other normal health care delivery sites. Through literature reviews and interviews with experts and professional associations, we identified four key components related to preparing for medical surge in a mass casualty event: (1) increasing hospital capacity, including beds, workforce, equipment, and supplies; (2) identifying and operating alternate care sites[5] when hospital capacity is overwhelmed; registering and credentialing volunteer medical professionals; and planning for appropriate altered standards of care[6] in order to save the most lives in a mass casualty event.

Federal and state entities both play roles in preparing for emergency preparedness. The Department of Homeland Security (DHS) has the overall federal responsibility under the Homeland Security Act of 2002 for managing national emergency preparedness.[7] In December 2006, the Congress passed the Pandemic and All-Hazards Preparedness Act (PAHPA). PAHPA designated the Secretary of Health and Human Services as the lead official for all federal public health and medical responses to public health emergencies, including medical surge.[8] Under the federal plan for responding to emergencies,[9] states have responsibility for producing emergency preparedness plans in coordination with regional and local entities, and both DHS and the Department of Health and Human Services (HHS) are responsible for supporting their efforts. In addition, the Department of Defense (DOD) and the Department of Veterans Affairs (VA) are expected to assist state and local entities in emergencies. A DOD directive authorizes local military hospitals to coordinate with state and local entities to plan for emergency preparedness, and DOD hospitals are authorized to accept civilian patients in a mass casualty event.[10] VA policies and procedures allow VA hospitals to participate in state and local emergency planning, and by statute VA may provide medical care to nonveterans in a mass casualty event.

As a result of the nation's need to prepare for potential terrorist attacks, naturally occurring disease outbreaks, or other natural disasters, members of the Congress asked that we undertake a study regarding the nation's preparedness for a mass casualty event. In this chapter, we examine the following questions: (1) What assistance has the federal government provided to help states prepare their regional and local health care systems for medical surge in a mass casualty event? (2) What have states done to prepare for medical surge in a mass casualty event? (3) What concerns have states identified as they prepare for medical surge in a mass casualty event?

To determine what assistance the federal government provided to states to help them prepare their regional and local health care systems for medical surge in a mass casualty event, particularly related to four key components of medical surge, we reviewed and analyzed national strategic planning documents. We also analyzed reports related to medical surge capacity issued by various entities, including the Agency for Healthcare Research and Quality (AHRQ), Centers for Disease Control and Prevention (CDC), Office of the Assistant Secretary for Preparedness and Response (ASPR), and the Joint Commission.[11] In addition, we obtained and reviewed documents from ASPR to determine the amount of funds awarded to states through its Hospital Preparedness Program's cooperative agreements. We also interviewed officials from ASPR, CDC, and DHS to identify and document criteria and guidance given to states to plan for medical surge. To determine what states have done to prepare for medical surge in a mass casualty event, we obtained and analyzed the 2006 and 2007 ASPR Hospital Preparedness Program cooperative agreement applications and 2006 midyear progress reports (the most current available information at the time of our data collection[12]) for the 50 states.[13]

We also reviewed the 15 sentinel indicators from these reports.[14] Although ASPR's 2006 guidance for these midyear progress reports did not provide specific criteria with which to evaluate recipients' performance on these sentinel indicators, we identified criteria to analyze the data provided for 5 of the indicators related to one of four key components—hospital capacity—from either ASPR's previous program guidance or DHS guidance.[15] (See app. I for a list of the sentinel indicators.) In addition, we obtained and reviewed 20 states' emergency preparedness planning documents relating to medical surge and interviewed officials from these states responsible for planning for medical surge. We selected the 20 states by identifying 2 states from each of the 10 HHS geographic regions—one with the most ASPR Hospital Preparedness Program funding and one with the least funding. These selection criteria allowed us to take into account population (program funding was awarded using a formula including, in part, population), geographic dispersion, and different geographic risk factors, such as the potential for hurricanes, tornadoes, or earthquakes. We obtained and reviewed DOD and VA policies and interviewed officials regarding their participation with state and local entities in emergency preparedness planning and response. To determine what concerns states identified as they prepared for medical surge, we interviewed emergency preparedness officials from the 20 states on their efforts related to four key components. We also asked what further assistance states might need from the federal government to help prepare their health care systems for medical surge. The information from these interviews is intended to provide a general description of what the 20 states have done to prepare for medical surge and is not generalizable to all 50 states. (See app. II for a more detailed scope and methodology.) We conducted our work from May 2007 through May 2008 in accordance with generally accepted government auditing standards.

Chapter 1

RESULTS IN BRIEF

The federal government has provided funding, guidance, and other assistance to help states prepare for medical surge in a mass casualty event. From fiscal years 2002 to 2007, the federal government awarded the states about $2.2 billion through ASPR's Hospital Preparedness Program to support activities to meet their preparedness priorities and goals, including medical surge. Further, the federal government developed, or contracted with experts to develop, guidance that was provided for states to use when preparing for medical surge—for example, DHS's *National Preparedness Guidelines* and *Target Capabilities List*. In addition, ASPR annually provided specific guidance for its Hospital Preparedness Program awardees on preparing for medical surge, including activities to assist states in following DHS's guidelines and meeting its targets. AHRQ issued guidance on *Reopening Shuttered Hospitals to Expand Surge Capacity*, which contains a checklist that states and local entities can use to identify organizations that could provide more resources during a medical surge.

In addition, ASPR project officers and CDC subject matter experts were available to provide assistance to states on issues related to medical surge.

Many states have made efforts related to three of the key components of medical surge, that is, increasing hospital capacity, planning for alternate care sites, and developing electronic medical volunteer registries, but fewer have addressed the fourth component, planning for altered standards of care. More than half of the 50 states had met or were close to meeting the criteria for the five medical-surge-related sentinel indicators for hospital capacity that we extracted from the Hospital Preparedness Program 2006 midyear progress reports. For example, 37 states reported that they could meet the criterion of being able to add enough beds to provide triage treatment and stabilization for at least 500 patients per million population within 24 hours of a mass casualty event, with another 4 states reporting that they could do so for from 400 to 499 patients per million population. In our 20-state review, we found that all were developing bed reporting systems and most were coordinating to various degrees with DOD and VA hospitals in an effort to expand their hospital capacity. Of the 20 states, 18 reported that they were in the process of selecting alternate care sites that used either fixed or mobile medical facilities. For example, one state had purchased three mobile medical facilities, each with 200 beds, to be located in different areas of the state. Additionally, 15 of the 20 states had begun registering volunteers in electronic medical volunteer registries, and 14 of those states reported that they had begun to verify the volunteers' medical qualifications, though few had conducted the verification at the level that most completely identified volunteers' skills and capabilities for providing care in a hospital. However, fewer of the states—7 of the 20—had adopted or were drafting altered standards of medical care to be used in response to a mass casualty event. For example, one state had prepared standards of care guidelines for the allocation of ventilators in that state and another state issued guidelines in February 2008 that call for suspending or relaxing state laws covering medical care and for allocating health care to save the most lives.

While the Hospital Preparedness Program has been operating since 2002, state officials in the 20 states we surveyed reported that they continued to face challenges in preparing for medical surge in a mass casualty event. They expressed concerns related to all four key components of medical surge. For example, some states reported that although they could increase numbers of hospital beds in a mass casualty event, they were concerned about staffing those beds because of current shortages in medical professionals, and some states reported concerns about reimbursement for medical services provided at alternate care sites. According to some state officials, volunteers were concerned that if state registries became part of a national database they might be required to provide services outside their own state. Some states reported that they had not begun work on altered standards of care guidelines, or had not completed drafting guidelines, because of the difficulty of addressing the medical, ethical, and legal issues involved in making life-or-death decisions in advance of a disaster about which patients would get or lose access to scarce resources. Finally, state officials also noted concerns related to other issues involved in preparing for medical surge, such as decreased federal funding for hospital emergency preparedness.

To further assist states in determining how they will allocate scarce medical resources in a mass casualty event, we recommend that the Secretary of HHS ensure that the department serve as a clearinghouse for sharing among the states altered standards of care guidelines that have been developed by individual states or medical experts. In commenting on a draft of this chapter, HHS was silent regarding our recommendation. HHS, DHS, DOD, and VA concurred with our findings.

Chapter 2

BACKGROUND

Federal responsibilities for assisting states in preparing for emergencies include developing national strategies, policies, and guidelines and providing funding to assist states in developing their emergency preparedness plans and programs. A critical element of emergency preparedness is preparing health care systems for medical surge in a mass casualty event, and consideration of hospital capacity, alternate care sites, electronic medical volunteer registries, and altered standards of care is key to this task.

FEDERAL RESPONSIBILITIES RELATING TO STATES' PREPAREDNESS FOR MEDICAL SURGE

DHS is responsible for developing national strategies, policies, and guidelines related to emergency preparedness. Additionally, DHS administers the Homeland Security Grant Program, which currently consists of four programs—the State Homeland Security Program, Urban Areas Security Initiative, Metropolitan Medical Response System, and Citizens Corps Program.[16] While these programs generally award funds to states and municipalities for the prevention and detection of terrorist acts, some funds can be spent on medical response, including medical surge activities.

HHS has the principal responsibility for helping states to prepare for medical surge. In December 2006, PAHPA established ASPR within HHS in order to enhance coordination of public health and medical surge. The act reauthorized and gave ASPR authority over the Hospital Preparedness Program,[17] which provides funds annually to 62 entities—the 50 states, 4 municipalities, 5 U.S. territories, and 3 Freely Associated States of the Pacific[18]—through cooperative agreements in order to strengthen their emergency readiness capabilities. Also, beginning in fiscal year 2009, HHS will require that states provide a 5 percent match to the amount of the federal cooperative agreement funding, through either state funds or in- kind contributions, such as office space or computer support for the program. In 2010 and subsequent years, the matching requirement will increase to 10 percent.

As part of the 2006 Hospital Preparedness Program, ASPR required all cooperative agreement recipients to submit midyear progress reports that include data on 15 sentinel indicators, 13 of which are related to medical surge. For example, one of the sentinel indicators is the number of hospitals that have the capacity to maintain at least one patient with a suspected highly infectious disease in a negative pressure isolation room.[19] PAHPA also gave ASPR authority for the Emergency System for Advance Registration of Volunteer Health Professionals (ESAR-VHP). ESAR-VHP supports state-based electronic databases designed to register health care personnel who volunteer to provide medical care in an emergency for the purpose of verifying their credentials. In order to continue to receive Hospital Preparedness Program funds, states must participate in ESAR-VHP by fiscal year 2009. Under PAHPA, HHS is required to link state electronic medical volunteer registries into a national registry.

DOD and VA do not have a federal responsibility in assisting states in planning and preparing for medical surge in a mass casualty event. However, since their hospitals are accredited by the Joint Commission, they are required to participate in at least one annual emergency preparedness exercise with their local community. In addition, because they are part of the local community, they would play a role in planning for and responding to local mass casualty events.

Key Components of Medical Surge in a Mass Casualty Event

According to *Homeland Security Presidential Directive 21 (HSPD-21) Public Health and Medical Preparedness*, issued in October 2007, mass casualty health care is a critical element of public health and medical preparedness. HSPD-21 is one of a series of executive orders released since September 11, 2001, establishing a national strategy to help protect the nation in the event of terrorist attacks or other catastrophic health events. It states that mass casualty health care capability needs to be different from "day-to-day" public health and medical operations, which "cannot meet the needs created by a catastrophic health event." It also states that the nation must develop a disaster medical capability that, among other things, is rapid, flexible, sustainable, integrated, and coordinated, and delivers appropriate treatment in the most ethical manner with available capabilities.

The four key components we identified follow:

- Hospital capacity: Following a mass casualty event, hospitals may need the ability to adequately care for a large number of additional patients. Strategies to increase hospital capacity include deferring elective procedures, applying more stringent triage for admitting patients, discharging patients early with follow-up by home health care personnel, and adding additional beds and equipment in areas of the hospital that are not normally used for inpatient care, such as outpatient examining rooms.
- Alternate care sites: A mass casualty event could overwhelm hospitals' capacity and require the establishment of alternate sites to provide health care services. Alternate care sites deliver medical care outside hospital settings for patients who would normally be treated as inpatients, and triage patients in order to sort those who need critical attention and immediate transport to the hospital from those with less serious injuries. In addition, alternate care sites manage unique considerations that might arise in the context of mass casualty events, including the delivery of chronic care; the distribution of vaccines; or the quarantine, grouping, or sequestration of patients potentially infected with an easily transmissible infectious disease. The development of alternate care sites involves several issues, including the level and scope of medical care to be delivered, the physical infrastructure required, staffing requirements for the delivery of such care, the medical equipment and supplies needed, and the management systems required to integrate such facilities with the overall delivery of health care. Additionally, there are two types of alternate care sites—fixed and mobile. Fixed facilities are nonmedical buildings that, because of their size or proximity to a hospital, can be adapted to provide medical care. Mobile medical facilities are either specialized units with surgical and intensive care capabilities that are based on tractor-trailer platforms or fully equipped hospitals stored in container systems that can be set up quickly.
- Electronic medical volunteer registries: In a time of emergency, it can be difficult for state and hospital officials who are organizing a response to use medical volunteers[20] unless they have been preregistered to determine

who is qualified to provide medical assistance. For example, immediately after the attacks on September 11, 2001, thousands of people spontaneously arrived in New York City to volunteer their assistance—many of whom volunteered to provide medical assistance to the victims of the attacks. However, authorities were unable to distinguish medically qualified from unqualified volunteers. Generally, an electronic medical volunteer registry would (1) preregister health care volunteers, (2) apply emergency credentialing standards to these registered volunteers, and (3) allow for the verification of the identity, credentials, and qualifications of registered volunteers in an emergency.[21]

Altered Standards of Care

In a mass casualty event, routine resource shortages would be significantly magnified and hospitals would have limited access to many needed resources, such as health care providers, equipment and supplies, and pharmaceuticals. As a result, it could be necessary to alter standards of medical care in a manner that is different from normal day-to-day circumstances and appropriate to the situation. For example, because of an influx of a large number of patients in a mass casualty event, adequate staffing of health care providers would be hindered by the current shortages of health care providers. Workforce shortages could result in hospitals changing their established standards of care, such as nurse-to-patient care ratios.[22]

Chapter 3

The Federal Government has Provided States with Funding, Guidance, and other Assistance to Prepare for Medical Surge

The federal government has provided funding, guidance, and other assistance to help states prepare their regional and local health care systems for medical surge in a mass casualty event. From fiscal years 2002 through 2007, the federal government awarded the states about $2.2 billion through ASPR's Hospital Preparedness Program to support activities to meet their preparedness priorities and goals, including medical surge. Further, the federal government developed, or contracted with experts to develop, guidance that was provided for states to use when preparing for medical surge. In addition, the federal government provided other assistance, such as conferences for states.

Funding to Prepare for Medical Surge

From fiscal years 2002 through 2007, HHS awarded states about $2.2 billion through ASPR's Hospital Preparedness Program[23] to support activities to strengthen their hospital emergency preparedness capabilities, including medical surge goals and priorities.[24] (See app. III for Hospital Preparedness Program cooperative agreement funding by state.) ASPR's 2007 Hospital Preparedness Program guidance specifically authorized states to use funds on activities such as the development of a fully operational electronic medical volunteer registry in accordance with ESAR-VHP guidance and the establishment of alternate care sites. We cannot report state-specific funding for four key components—hospital capacity, alternate care sites, electronic medical volunteer registries, and altered standards of care—because state expenditure reports did not disaggregate the dollar amount spent on specific activities related to these components. During fiscal years 2003 through 2007, DHS's Homeland Security Grant Program also awarded the states funds that were used for a broad variety of emergency preparedness activities and may have included medical surge activities. However, most of these DHS grant funds were not targeted to medical surge activities, and states do not report the dollar amounts spent on these activities.

GUIDANCE TO PREPARE FOR MEDICAL SURGE

The federal government developed, or contracted with experts to develop, guidance for states to use in preparing for medical surge. DHS developed overarching guidance, including the *National Preparedness Guidelines* and the *Target Capabilities List*. The *National Preparedness Guidelines* describes the tasks needed to prepare for a medical surge response to a mass casualty event, such as a bioterrorist event or natural disaster, and establishes readiness priorities, targets, and metrics to align the efforts of federal, state, local, tribal, private-sector, and nongovernmental entities. The *Target Capabilities List* provides guidance on building and maintaining capabilities, such as medical surge, that support the *National Preparedness Guidelines*. The medical surge capability includes activities and critical tasks needed to rapidly and appropriately care for the injured and ill from mass casualty events and to ensure that continuity of care is maintained for non-incident-related injuries or illnesses.[25]

In addition, ASPR provided states with specific guidance related to preparing for medical surge in a mass casualty event, including annual guidance for its Hospital Preparedness Program cooperative agreements, guidance for developing ESAR-VHP-compliant electronic medical volunteer registries, and guidance to develop a hospital bed tracking system. The Hospital Preparedness Program cooperative agreement guidance included activities to assist states in following DHS's guidelines and meeting its targets. ASPR's ESAR-VHP guidelines provide states with common definitions, standards, and protocols, which can aid in forming a national network to facilitate the deployment of medical volunteers for any emergency among states. For example, ESAR-VHP registration guidelines categorize medical volunteers by profession, ranging from physicians to mental health counselors. ESAR-VHP guidelines also include four different levels of credentialing based on verification of each volunteer's qualifications.[26] ASPR provided guidance to states for the Hospital Available Beds for Emergencies and Disasters (HAvBED) system, which is an inpatient bed tracking system designed to allow emergency response entities to know where and what type of additional hospital beds are available, in order to know which hospitals still have capacity to receive patients. HAvBED reports the number of beds vacant/available at the aggregate state level to HHS. To enhance consistency among state- reported data, HAvBED provides standard definitions of beds and data elements each system must incorporate when reporting bed availability during a mass casualty event.[27]

Additionally, HHS worked through AHRQ and contracted with nonfederal entities to develop publications for states to use when preparing for medical surge. For example, AHRQ published the document *Mass Medical Care with Scarce Resources: A Community Planning Guide* to provide states with information that would help them in their efforts to prepare for medical surge, such as specific circumstances they may face in a mass casualty event. This publication notes that the state may be faced with allocating medical resources during a mass casualty event, such as determining which patients will have access to mechanical ventilation. The publication recommends that the states develop decision-making guidelines on how to allocate these medical resources. The RAND Corporation developed the publication *Learning from Experience: The Public Health Response to West Nile Virus, SARS, Monkeypox, and Hepatitis A Outbreaks in the United States*, which provides states with information on challenges that they may face in a disease outbreak or bioterrorist attack.[28] AHRQ also published *Reopening Shuttered Hospitals to Expand Surge Capacity*, which contains an action checklist that can be used by states and local entities to identify organizations that have an interest or responsibility in preparing for medical surge, and to determine what resources each could provide. (See app. III for a list of federal guidance.)

Other Federal Assistance to Prepare for Medical Surge

To support states' efforts to prepare for medical surge, the federal government also provided other assistance such as conferences and electronic bulletin boards for states to use in preparing for medical surge. States were required to attend annual conferences for Hospital Preparedness Program cooperative agreement recipients, where ASPR provided forums for discussion of medical surge issues.

(See app. III for a list of federal conferences.) Additionally, ASPR's Web site contained links to related published documents, and states were given access to an ASPRoperated electronic bulletin board to communicate with other states on medical surge issues related to the Hospital Preparedness Program. Furthermore, ASPR project officers and CDC subject matter experts were available to provide assistance to states on issues related to medical surge. For example, CDC's Division of Healthcare Quality Promotion developed cross-sector workshops for local communities to bring their emergency management, medical, and public health officials together to focus on emergency planning issues, such as developing alternate care sites.

Chapter 4

MANY STATES HAVE MADE EFFORTS TO INCREASE HOSPITAL CAPACITY, PLAN FOR ALTERNATE CARE SITES, AND DEVELOP ELECTRONIC MEDICAL VOLUNTEER REGISTRIES, BUT FEWER HAVE PLANNED FOR ALTERED STANDARDS OF CARE

Many states have made efforts related to three of the key components for preparing for medical surge, that is, increasing hospital capacity, planning for alternate care sites, and developing electronic medical volunteer registries, but fewer have implemented the fourth, planning for altered standards of care. More than half of the 50 states were meeting or close to meeting the criteria for the five medical-surge-related sentinel indicators for hospital capacity. In our 20-state review, we found that all were developing bed reporting systems and almost all of the states with DOD and VA hospitals were engaging in various levels of coordination with those hospitals in an effort to expand their hospital capacity. Of the 20 states, 18 reported that they were in the process of selecting alternate care sites that used either fixed or mobile medical facilities. Additionally, 15 of the 20 states had begun registering volunteers in electronic medical volunteer registries. However, only 7 of the 20 states had adopted or were drafting altered standards of care for specific medical interventions to be used in response to a mass casualty event.

ALL STATES WERE MAKING EFFORTS TO EXPAND HOSPITAL CAPACITY

More than half of the states met or were close to meeting the criteria for the five surge-related sentinel indicators for hospital capacity that we reviewed from the Hospital Preparedness Program 2006 midyear progress reports,[29] the most recent available data at the time of our analysis.[30] (See table 1 for the five sentinel indicators and the associated criteria.) Twenty- four of the states reported that all of their hospitals were participating in the state's program funded by the ASPR Hospital Preparedness Program, with another 14 states reporting that 90 percent or more of their hospitals were participating. Forty-three of the 50 states have increased their hospital capacity by ensuring that at least one health care facility in each defined region could support initial evaluation and treatment of at least 10 patients at a time (adult and pediatric) in negative pressure isolation within 3 hours of an event. Regarding individual hospitals' isolation capabilities, 32 of the 50 states met the requirement that all hospitals in the state that participate in the Hospital Preparedness Program be able to maintain at least one suspected highly infectious disease case in negative pressure isolation; another 10 states had that capability in 90 to 99 percent of their participating hospitals. Thirty-seven of the 50 states reported meeting the criteria that within 24 hours of a mass casualty event, their hospitals would be able to add enough beds to provide triage treatment and stabilization for another 500 patients per million population; another 4 states reported that their hospitals could add enough beds for from 400 to 499 patients per million population. Finally, 20 states reported that all their participating hospitals had access to pharmaceutical caches that were sufficient to cover hospital personnel (medical and ancillary), hospital- based emergency first responders, and family members associated with their facilities for a 72-hour period; another 6 states reported that from 90 to 99 percent of their participating hospitals had sufficient pharmaceutical caches. (See app. IV for further information.)

In our further review of 20 states, all 20 states reported that they had developed or were developing bed reporting systems to track their hospital capacity—the first of four key components related to preparing for medical surge. Eighteen of the 20 states reported that they had systems in place that could report the number of available hospital beds within the state. All 18 of these states reported that their systems met ASPR HAvBED standards.[31] For example, in early 2005 one state completed development of a statewide Web-based bed tracking system designed to track the emergency status of all health care facilities.[32] The system has the capacity to present information by individual facility as well as by county. The 2 states that reported that they did not have a system that could meet HAvBED requirements said that they would meet the requirements by August 8, 2008.[33]

Our review also found that of the 10 states with DOD hospitals, 9 reported coordinating with DOD hospitals to plan for emergency preparedness and increase hospital capacity. For example, in one state DOD hospital officials served on state-level emergency preparedness committees and participated in training and exercises. The remaining state said it could not report whether the DOD hospitals participated in such activities because these activities were coordinated at the local level. Eight of the 10 states also reported that DOD hospitals in their state would accept civilian patients in the event of a mass casualty event if resources were available.[34] The 2 remaining states did not know whether their DOD hospitals would accept civilian patients, although one of these states said that there had been discussions about this possibility between the state and DOD.

Of the 19 states that have VA hospitals, all reported that at least some of the VA hospitals took part in the states' hospital preparedness programs or were included in planning and exercises for medical surge.[35] For example, VA hospitals in one state were participating in state, regional, and local planning for emergency preparedness along with other hospitals in an effort to increase surge capacity and come closer to the state's goal of 500 beds for every 1 million population, a VA official said. In another state, a VA hospital was planning with state emergency preparedness officials and DOD hospitals to prepare for any mass casualty event that could occur during a major public event taking place in the state later that year. VA officials stated that individual hospitals cannot precommit resources— specific numbers of beds and assets—for planning purposes, but can accept nonveteran patients and provide personnel, equipment, and supplies on a case-by-case basis during a mass casualty event.[36] Twelve of the 19 states reported that VA hospitals would accept or were likely to accept nonveteran patients in the event of a medical surge if space were available and veterans' needs had been met. Four of the 19 states reported that their VA hospitals would not accept nonveteran patients in the event of a medical surge, 2 states reported that they did not know if the VA hospitals would accept nonveteran patients, and 1 state reported that some of its VA hospitals would take nonveteran patients and others would not.

Table 1. Medical-Surge-Related Sentinel Indicators for Hospital Capacity from Hospital Preparedness Program 2006 Midyear Progress Reports and Our Associated Criteria

Sentinel indicator	Criteria we used[a]
Hospital participation: Total number of participating hospitals statewide.	State reported that 100 percent of its hospitals were participating in state hospital preparedness programs supported by ASPR funding.
Regional negative pressure isolation:[b] Number of states' defined regions that have regional facilities to support the initial evaluation and treatment of at least 10 adult and pediatric patients at a time in negative pressure isolation within 3 hours post-event.	State could identify at least one health care facility in each definedsubstate region[c] that could support initial evaluation and treatment of at least 10 patients (adult and pediatric) at a time in negative pressure isolation within 3 hours of an event.
Hospital negative pressure isolation: Number of participating hospitals statewide that have the capacity to maintain at least one suspected highly infectious disease case in negative pressure isolation.	State reported that all participating hospitals in the state were ableto maintain at least one suspected highly infectious disease case in negative pressure isolation.
Surge beds: Number of beds statewide, above the current daily staffed bed capacity, that the state is capable of surging beyond within 24 hours post-event.	State reported that within 24 hours of a mass casualty event, its hospitals would be able to add enough additional hospital beds to the state's current daily staffed bed capacity to provide triage treatment and initial stabilization for an additional 500 patients per million population.
Pharmaceutical caches: Number of participating hospitals statewide that have access to pharmaceutical caches sufficient to cover hospital personnel (medical and ancillary), hospital-based emergency first responders, and family members associated with their facilities for a 72-hour period.	State reported that all its participating hospitals had access to pharmaceutical caches that were sufficient to cover hospital personnel (medical and ancillary), hospital-based emergency first responders, and family members associated with its facilities for a 72-hour period.[d]

Source: GAO analysis of recipient-reported data, the Health Resources and Services Administration's (HRSA) 2005 National Bioterrorism Hospital Preparedness Program guidance, and DHS's *Target Capability List*.

Note: Prior to March 2007, the Hospital Preparedness Program was administered by HHS's HRSA and was named the National Bioterrorism Hospital Preparedness Program.

[a] Because ASPR's guidance for the 2006 midyear progress reports did not provide specific criteria with which to evaluate recipients' performance on these sentinel indicators, we identified criteria to analyze the data provided for five of the indicators related to hospital capacity from either ASPR's previous program guidance or DHS's *Target Capabilities List.*

[b] Negative pressure isolation rooms maintain a flow of air into the room to ensure that contaminants and pathogens cannot escape from the room to other parts of the facility and to protect the health of workers and other patients.

[c] Each recipient was required to subdivide its state into regions.

[d] Off icials from one state said they did not know how to determine whether their hospitals had access to caches that were "sufficient."

In planning to increase hospital capacity, most of the 20 states we surveyed reported that they used federal guidance and technical assistance. Eleven states reported that they used ASPR's Hospital Preparedness Program cooperative agreement guidance, and 9 states used ASPR's *Medical Surge Capacity and Capabilities Handbook.* Three states also reported that they used CDC's Public Health Emergency Preparedness Program cooperative agreement guidance. In addition, 2 states reported that they consulted with ASPR project officers when planning for hospital capacity.

EIGHTEEN STATES WERE SELECTING ALTERNATE CARE SITES

Eighteen of the 20 states reported that they were in the process of selecting alternate care sites, and the 2 remaining states reported that they were in the early planning stages in determining how to select sites. Of the 18 states, 10 reported that they had also developed plans for equipping and staffing some of the sites. For example, one state had developed standards and guidance for counties to use when implementing fixed alternate care sites and had stockpiled supplies and equipment for these sites. The counties were responsible for identifying and operating these sites. According to state officials, while most counties were still identifying fixed sites, some counties had established memorandums of understanding with various facilities, including churches, schools, military facilities, and shopping malls. In addition, the state purchased three state-run mobile medical facilities, each with 200 beds, which were stored in the northern, central, and southern parts of the state. Another state, which expects significant transportation difficulties during a natural disaster, had acquired six mobile medical tent facilities of either 20 or 50 beds that were stored at hospital facilities across the state. This state also planned to identify fixed facility alternate care sites, which would provide medical services to people who could not take care of themselves at home but did not need to be in a hospital. Each of these fixed sites was expected to serve 1,000 casualties. One of the 2 states that were in the early planning stages was helping local communities formalize site selection agreements, and the second state had drafted guidance for alternate care sites that was expected to be released early in 2008.

Most states reported using AHRQ guidance when planning for alternate care sites. For example, 18 states reported that they used AHRQ's guidance, such as *Rocky Mountain Regional Care Model for Bioterrorist Events, Alternate Care Site Selection Tool*, and *Reopening Shuttered Hospitals to Expand Surge Capacity*. A few states used other federal guidance, such as DHS's National Incident Management System and National Disaster Management System guidance, when planning alternate care sites. Five states also reported that they used DOD guidance when planning alternate care sites, including DOD's *Modular Emergency Medical System*.[37]

FIFTEEN STATES HAD BEGUN REGISTERING VOLUNTEERS IN ELECTRONIC MEDICAL VOLUNTEER REGISTRIES

Fifteen of the 20 states reported that they had begun registering medical volunteers and identifying their medical professions in an electronic registry, and the remaining 5 states were developing their electronic registries and had not registered any volunteers. For 2006, ESAR-VHP guidance identified seven categories of health care professionals ranging from physicians to mental health counselors that should be included in the states' registries.[38] Of the 15 states that reported that they had begun registering volunteers, 3 states had registered volunteers in more than eight categories, 3 states had registered volunteers in five to seven categories, and the remaining 9 states had registered volunteers in four or fewer categories, often concentrating on nurses. Officials from 4 of the 5 remaining states that had not begun registering volunteers reported that they anticipated registering volunteers by the spring or summer of 2008. An official from the other state reported that state officials did not know when they would begin to register volunteers.

Of the 15 states that reported they were registering volunteers, 12 reported they had begun to verify the volunteers' medical qualifications, though few had conducted the verification to assign volunteers to the highest level, Level 1. If a volunteer is assigned to Level 4, it means that the state has not verified any medical qualifications, such as licenses or certifications in medical subspecialties. Three of the 15 states had registered volunteers solely at Level 4. Seven of the 12 states had credentialed some volunteers no higher than Level 3, meaning they had verified the licenses of some of the volunteers. For example, one state had verified the credentials and assigned all of its 1,498 registered volunteers at Level 3. Another 3 of the 12 states had assigned volunteers to no higher than Level 2, meaning these states had conducted additional verification of medical qualifications, such as degrees. For example, one state had assigned its registered volunteer nurses at Level 2. The remaining 2 states had assigned a small number of volunteers at Level 1. For example, one state had assigned 2 of 955 volunteers at Level 1. At Level 1, all of a volunteer's medical qualifications, which identify their skills and capabilities, have been verified and the volunteer is ready to provide care in any setting, including a hospital.

Nineteen of the 20 states reported that they used ASPR's *ESAR-VHP Interim Technical and Policy Guidelines, Standards, and Definitions* when developing registries. Eight of the 20 states also reported that they used information obtained from the annual ESAR-VHP conferences to help develop their volunteer medical registry systems.

Seven of the 20 States Were Planning for Altered Standards of Medical Care

In our 20-state review of efforts related to the fourth key component, we found that 7 states had adopted or were drafting altered standards of care for specific medical issues. Three of the 7 states had adopted some altered standards of care guidelines. For example, one state had prepared a standard of care for the allocation of ventilators in an avian influenza pandemic, which one state official reported would also be applicable during other types of emergencies.[39] Another state issued guidelines in February 2008 for allocating scarce medical resources in a mass casualty event that call for suspending or relaxing state laws covering medical care and for explicit rationing of health care to save the most lives, and require that the same allocation guidelines be used across the state. For example, during a mass casualty event in this state, hospitals could ignore their nurse-patient ratios and nurses could be assigned to jobs outside their specific area of expertise. In addition, nonlicensed individuals, or retired health care providers whose licenses had lapsed, could be recruited to provide emergency care. For example, a nonmedical hospital employee who had experience as a military medic could get an emergency credential to stitch up wounds or start intravenous lines. According to an official, the state had not completed all of the guidelines for allocation of scarce resources that it planned to develop. The state recently convened a panel of ethicists and providers to address which specific categories of patients would receive scarce resources, such as vaccines and ventilators, when shortages existed.

Of the 13 states that had not adopted or drafted altered standards of care, 11 states were beginning discussions with state stakeholders, such as medical professionals and lawyers, related to altered standards of care, and 2 states had not addressed the issue. One state reported that its state health department planned to establish an ethics advisory board to begin discussion on altered standards of care guidelines. Another state had developed a "white paper" discussing the need for an altered standards of care initiative and planned to fund a symposium to discuss this initiative.

Six of the seven states that had adopted or were drafting altered standards of care guidelines reported using AHRQ documents, such as *Altered Standards of Care in Mass Casualty Events* and *Mass Medical Care with Scarce Resources: A Community Planning Guide*. Officials from one state reported that they had also used CDC documents and the federal government's pandemic influenza Web site[40] when planning for altered standards of care.

Chapter 5

STATES REPORTED CONCERNS RELATED TO ALL FOUR KEY COMPONENTS WHEN PREPARING FOR MEDICAL SURGE

While the Hospital Preparedness Program has been operating since 2002, state officials in the 20 states we surveyed reported that they faced continuing challenges in preparing for medical surge in a mass casualty event. Even though many states have made efforts to increase hospital capacity, provide care at alternate care sites, identify and use medical volunteers, and develop appropriate altered standards of care, they expressed concerns related to all four of these key components of medical surge. State officials also noted concerns related to programmatic and regulatory issues involved in preparing for medical surge in a mass casualty event.

HOSPITAL CAPACITY CONCERNS

State officials raised several concerns related to their ability to increase hospital capacity, including maintaining adequate staffing levels during mass casualty events, a problem that was more acute in rural communities. While 19 of 20 states we surveyed reported that they could increase numbers of hospital beds in a mass casualty event,[41] some state officials were concerned about staffing these beds because of current shortages in medical professionals, including nurses and physicians. Some state officials reported that their states faced problems in increasing hospital capacity because many of their rural areas had no hospital or small numbers of medical providers. For example, officials from a largely rural state reported that in many of the state's medically underserved areas hospitals currently have vacant beds because they cannot hire medical professionals to staff them. In addition, these officials reported that because their hospitals did not provide pediatric intensive care or burn care services and instead transferred these patients to neighboring states, the state might not be able to provide these services during a mass casualty event.

State officials also reported that as time passed and no mass casualty events occurred, increasing hospital capacity for a mass casualty event seemed to be a waning priority for hospital chief executive officers. State officials reported that it was difficult to continue to engage private-sector hospital chief executive officers in emergency preparedness activities at a time when these hospitals were facing day-to-day financial problems. For example, officials from one state reported that hospitals in the state were consolidating and closing, and officials from another state reported that fewer hospitals were applying for ASPR Hospital Preparedness Program funds. Officials from two other states reported that progress in preparing emergency plans had slowed, especially for the smaller rural facilities, because the Hospital Preparedness Program allows states to use these funds to hire staff to assist with emergency planning but prohibits hospitals from doing so. According to officials from one of these states, hospital staff have had limited time to spend on emergency planning activities because they must first attend to the operational needs of the hospital.

ALTERNATE CARE SITE CONCERNS

Some state officials reported that it was difficult to identify appropriate fixed facilities for alternate care sites. Officials from two states reported that some small, rural communities had few facilities that would be large enough to house an alternate care site. Officials from some states also reported that some of the facilities that could be used as alternate care sites had already been allocated for other emergency uses, such as emergency shelters.

State officials also reported concerns about reimbursement for medical services provided at alternate care sites, which are not accredited health care facilities. During the response to Hurricane Katrina, the Secretary of HHS waived a number of statutory and regulatory requirements related to medical care, and this waiver allowed for reimbursement of medical care provided in alternate care sites.[42] However, officials from several states said that hospitals would prefer to know ahead of time under what circumstances they would receive reimbursement from the Centers for Medicare & Medicaid Services (CMS) for medical care provided in alternate care sites during a mass casualty event. State officials said that having such information would make planning and exercising easier and more realistic. CMS officials told us it would be very difficult to provide specific guidance that would apply to all medical surge events and that the agency preferred to issue guidance on a case-by-case basis following visits to alternate care sites by CMS or Joint Commission officials during the emergency.[43] For example, after Hurricane Katrina, CMS officials visited alternate care sites and the Secretary of HHS relaxed reimbursement requirements for medical care provided in a hospital parking lot, the convention center, and a department store.

State officials also told us they were unclear how certain federal laws and regulations that relate to medical care — specifically, the privacy rule issued by HHS under the Health Insurance Portability and Accountability Act of 1996 (HIPAA)[44] and the Emergency Medical Treatment and Labor Act (EMTALA)[45] — would apply in a mass casualty event, especially if the care were provided in an alternate care site and not a hospital. EMTALA requires hospital emergency rooms at Medicare-participating hospitals to screen and treat for emergency medical conditions all individuals who seek treatment. The HIPAA privacy rule prohibits the unauthorized disclosure of individually identifiable health information by health care providers and certain other entities.[46] The Social Security Act authorizes the Secretary of HHS to waive EMTALA and certain requirements under the HIPAA privacy rule during national emergencies, such as a mass casualty event.[47] Federal guidance published in 2006 describes circumstances where provisions related to emergency treatment and privacy protections were temporarily suspended. AHRQ's publication *Providing Mass Medical Care with Scarce Resources: A Community Planning Guide* states that requiring hospitals to adhere to EMTALA requirements during a mass casualty event could be unworkable because of the large number of casualties. It notes that during Hurricane Katrina, HHS temporarily suspended the application of EMTALA in affected regions. This allowed hospitals to provide individuals' medical screening examination at, or transfer them to, alternate care sites, such as a convention center and department store. During Hurricane Katrina, HHS also temporarily relaxed the sanctions and penalties arising from noncompliance with certain provisions of the HIPAA privacy rule, including the requirements to obtain a patient's agreement to speak with family members or friends. HHS provided details of these waivers on its Hurricane Katrina Web site.[48]

ELECTRONIC MEDICAL VOLUNTEER REGISTRY CONCERNS

Some states reported that medical volunteers might be reluctant to join a state electronic medical volunteer registry if it is used to create a national medical volunteer registry. PAHPA requires ASPR to use the state-based registries to create a national database. According to state officials, some volunteers do not want to be part of a national database because they are concerned that they might be required to provide services outside their own state. Officials from one state reported that since PAHPA was enacted, recruiting of medical volunteers was more difficult and that the federal government should clarify whether national deployment is a possibility. ASPR officials said that they would not deploy medical volunteers nationally without working through the states.

Finally, some states expressed concerns about coordination among programs that recruit medical volunteers for emergency response. Officials from one state reported that federal volunteer registration requirements for the Medical Reserve Corps (MRC)[49] and the ESAR-VHP programs had not been coordinated, resulting in duplication of effort for volunteers. For example, the volunteers registered in the MRC units in that state also were expected by the state to register in the state electronic medical volunteer registry. Officials from a second state reported that a volunteer for one program that recruits medical volunteers is often a potential volunteer for another such program, which could result in volunteers being double-counted. For example, an emergency medical technician registered in the electronic medical volunteer registry may also volunteer for an MRC unit, a Disaster Medical Assistance Team (DMAT),[50] and the American Red Cross. This may cause staffing problems in the event of an emergency when more than one volunteer program is activated.

ALTERED STANDARDS OF CARE CONCERNS

Some state officials reported that they had not begun work on altered standards of care guidelines, or had not completed drafting guidelines, because of the difficulty of addressing the medical, ethical, and legal issues involved. For example, HHS estimates that in a severe influenza pandemic almost 10 million people would require hospitalization,[51] which would exceed the current capacity of U.S. hospitals and necessitate difficult choices regarding rationing of resources.[52] HHS also estimates that almost 1.5 million of these people would require care in an intensive care unit and about 740,000 people would require mechanical ventilation. Even with additional stockpiles of ventilators, there would likely not be a sufficient supply to meet the need. Since some patients could not be put on a ventilator, and others would be removed from the ventilator, standards of care would have to be altered and providers would need to determine which patients would receive them. In addition, some state officials reported that medical volunteers are concerned about liability issues in a mass casualty event. Specifically, state officials reported that hospitals and medical providers might be reluctant to provide care during a mass casualty event, when resources would be scarce and not all patients would be able to receive care consistent with established standards. According to these officials, these providers could be subject to liability if decisions they made about altering standards of care resulted in negative outcomes. For example, allowing staff to work outside the scope of their practice, such as allowing nurses to diagnose and write medical orders, could place these individuals at risk of liability.

While some states reported using AHRQ's *Mass Medical Care with Scarce Resources: A Community Planning Guide* to assist them as they developed altered standards of care guidelines, some states also reported that they needed additional assistance. States said that to develop altered standards of care guidelines they must conduct activities such as collecting and reviewing published guidance and convening experts to discuss how to address the medical, ethical, and legal issues that could arise during a mass casualty event. Four states reported that, when developing their own guidelines on the allocation of ventilators, they were using guidance from another state. This state estimated that a severe influenza pandemic would require nearly nine times the state's current capacity for intensive care beds and almost three times its current ventilator capacity, which would require the state to address the rationing of ventilators. In March 2006 the state convened a workgroup to consider clinical and ethical issues in the allocation of mechanical ventilators in an influenza pandemic.[53] The state issued guidelines on the rationing of ventilators that include both a process and an evaluation tool to determine which patients should receive mechanical ventilation. The guidelines note that the application of this process and evaluation tool could result in withdrawing a ventilator from one patient to give it to another who is more likely to survive—a scenario that does not explicitly exist under established standards of care. Additionally, some states suggested that the federal government could help their efforts in several ways, such as by convening medical, public health, and legal experts to address the complex issues associated with allocating scarce resources during a mass casualty event, or by developing demonstration projects to reveal best practices employed by the various states. Recently, the Task Force for Mass

Critical Care, consisting of medical experts from both the public and the private sectors, provided guidelines for allocating scarce critical care resources in a mass casualty event that have the potential to assist states in drafting their own guidelines. The task force's guidelines, which were published in a medical journal in May 2008,[54] provide a process for triaging patients that includes three components—inclusion criteria, exclusion criteria, and prioritization of care. The exclusion criteria include patients with a high risk of death, little likelihood of long-term survival, and a corresponding low likelihood of benefit from critical care resources. When patients meet the exclusion criteria, critical care resources may be reallocated to patients more likely to survive.

Other Programmatic and Regulatory Concerns

Many state officials raised concerns about other federal programmatic and regulatory challenges, such as program funding cycles, decreased federal funding for hospital emergency preparedness, and new requirements for state matching funds. State officials reported that ASPR's Hospital Preparedness Program's single-year funding cycles had made planning and operating state emergency preparedness programs challenging, in part because it is difficult to plan and implement program activities in a single year. One state official suggested that using a 3-year funding cycle for the Hospital Preparedness Program would allow for long-term planning with more realistic work plans. It would also allow for more time for program development and less time spent on program administration. ASPR officials said that they were aware of the concern and were considering a transition to a multiyear funding cycle beginning in 2009. Another concern expressed by some state officials was that federal funding for ASPR's Hospital Preparedness Program had decreased while program requirements had increased, making it difficult for states to plan for maintenance of emergency preparedness systems, meet new requirements, and replace expired supplies. Hospital Preparedness Program funds decreased about 18 percent from fiscal year 2004 to fiscal year 2007. Finally, many state officials were concerned about the new requirement for matching funds. Beginning in fiscal year 2009, states that want to receive ASPR's Hospital Preparedness Program funds will have to match 5 percent of the federal funds with either state funds or in-kind contributions.

CONCLUSION

Though states have begun planning for medical surge in a mass casualty event, only 3 of the 20 states in our review have developed and adopted guidelines for using altered standards of care. HHS has provided broad guidance that establishes a framework and principles for states to use when developing their specific guidelines for altered standards of care. However, because of the difficulty in addressing the related medical, ethical, and legal issues, many states are only beginning to develop such guidelines for use when there are not enough resources, such as ventilators, to care for all affected patients. In a mass casualty event, such guidelines would be a critical resource for medical providers who may have to make repeated life-or-death decisions about which patients get or lose access to these resources—decisions that are not typically made in routine circumstances. Additionally, these guidelines could help address medical providers' concerns about ethics and liability that may ensue when negative outcomes are associated with their decisions. In its role of assisting states' efforts to plan for medical surge, HHS has not collected altered standards of care guidelines that some states and medical experts have developed and made them available to other states. Once a mass casualty event occurs, difficult choices will have to be made, and the more fully the issues raised by such choices are discussed prior to making them, the greater the potential for the choices to be ethically sound and generally accepted.

RECOMMENDATION FOR EXECUTIVE ACTION

To further assist states in determining how they will allocate scarce medical resources in a mass casualty event, we recommend that the Secretary of HHS ensure that the department serve as a clearinghouse for sharing among the states altered standards of care guidelines that have been developed by individual states or medical experts.

AGENCY COMMENTS AND OUR EVALUATION

We requested comments on a draft of this chapter from HHS, DHS, DOD, and VA. These agencies' comments are reprinted in appendixes V, VI, VII, and VIII, respectively.

In commenting on this draft, HHS said our chapter was a fair representation of the progress that has been made to improve medical surge capacity. HHS was silent regarding our recommendation that the department serve as a clearinghouse for sharing among the states altered standards of care guidelines developed by individual states or medical experts. HHS provided technical comments, which we incorporated where appropriate.

In commenting on this draft, DHS concurred with our findings and raised two issues. With regard to the phrase "altered standards of care," DHS said that the definition of standard of care implies that the standard does not change but "rather it is the type, or level, of care that is altered," and that this distinction highlights the need to prepare the public "for a different look to health care" in a mass casualty incident. We agree that efforts to inform the public would be beneficial because of the need for enhanced public awareness about how medical care might be delivered in an emergency, but our chapter focused on addressing states' concerns about the medical, ethical, and legal issues involved in drafting altered standards of care guidelines. DHS also characterized our recommendation as calling for "passive guidance" and suggested that HHS may need to explore the possibility of producing guidance to direct states' discussion on rationing of scarce resources. However, we believe a clearinghouse role is more appropriate for HHS than a directive role because the delivery of medical care is a state, local, and private function.

DOD concurred with our findings and conclusions. VA concurred with our findings and said that inconsistencies from state to state regarding VA medical centers' stance toward treating nonveterans in an emergency stem from the centers' varying capabilities to provide emergency medical treatment. VA said, for example, that not all medical centers provide emergency services or have the same level of emergency supplies. Nevertheless, VA confirmed its authority to provide care in emergency situations and specifically acknowledged that it is authorized to provide emergency care to nonveterans on a humanitarian basis. Finally, VA also highlighted its federal role in responding to disasters under Emergency Support Function #8, the Robert T. Stafford Disaster Relief and Emergency Assistance Act, and the National Response Framework, which was beyond the scope of our chapter.

Cynthia A. Bascetta
Director, Health Care

LIST OF REQUESTERS

The Honorable Judd Gregg Ranking Member
Committee on the Budget United States Senate
The Honorable Charles E. Grassley Ranking Member
Committee on Finance
United States Senate
The Honorable Bennie G. Thompson Chairman
Committee on Homeland Security House of Representatives
The Honorable Henry A. Waxman
Chairman
Committee on Oversight and Government Reform House of Representatives
The Honorable Edward J. Markey House of Representatives

APPENDIX I: FIFTEEN HOSPITAL PREPAREDNESS PROGRAM 2006 SENTINEL INDICATORS

Sentinel indicator[a]

1. he number of hospitals statewide.
2. Total number of participating hospitals statewide.
3. Total population statewide.
4. Number of beds statewide, above the current daily staffed bed capacity, that awardee is capable of surging beyond within 24 hours post-event.
5. Number of participating hospitals statewide that have access to pharmaceutical caches sufficient to cover hospital personnel (medical and ancillary), hospital-based emergency first responders and family members associated with their facilities for a 72-hour period.
6. Number of participating hospitals statewide that have the capacity to maintain at least one suspected highly infectious disease case in negative pressure isolation.
7. Number of awardees' defined regions that have regional facilities to support the initial evaluation and treatment of at least 10 adult and pediatric patients at a time in negative pressure isolation within 3 hours post-event.
8. Number of ambulatory and nonambulatory persons that can be decontaminated within a 3-hour period, statewide.
9. Number of health care personnel, statewide, trained through competency-based programs.

Table of Appendix 1. Continued

10. Number of hospital lab personnel, statewide, trained in the protocols for referral of clinical samples and associated information.
11. Functional state-based ESAR-VHP[b] system in place that allows qualified, competent volunteer health care professionals to register for work in hospitals or other facilities during an emergency situation.
12. Number of volunteer health professionals by discipline and credentialing level currently registered in the state-based ESAR-VHP[b] system.
13. Number of drills conducted during the fiscal year 2006 budget period that included hospital personnel, equipment, or facilities.
14. Number of tabletop exercises conducted during the fiscal year 2006 budget period that included hospital personnel, equipment, or facilities.
15. Number of functional exercises conducted during the fiscal year 2006 budget period that included hospital personnel, equipment, or facilities.

Source: Office of the Assistant Secretary for Preparedness and Response (ASPR).

[a] The five sentinel indicators that were analyzed in this report for hospital capacity are 2, 4, 5, 6, and 7.

[b] ESAR-VHP is the Emergency System for Advance Registration of Volunteer Health Professionals.

Appendix II: Scope and Methodology

To determine what assistance the federal government has provided to help states prepare their regional and local health care systems for medical surge in a mass casualty event, particularly related to four key components—hospital capacity, alternate care sites, electronic medical volunteer registries, and altered standards of care—we reviewed and analyzed national strategic planning documents and identified links among federal policy documents on emergency preparedness. We also reviewed and analyzed studies and reports related to medical surge capacity issued by the Congressional Research Service, the Department of Health and Human Services' (HHS) Office of Inspector General, the Agency for Healthcare Research and Quality (AHRQ), the Centers for Disease Control and Prevention (CDC), the Office of the Assistant Secretary for Preparedness and Response (ASPR), the Joint Commission,[1] and other experts. In addition, we obtained and reviewed documents from ASPR to determine the amount of funds awarded to states through its Hospital Preparedness Program's cooperative agreements.[2] We did not review funding documents from the Department of Homeland Security's (DHS) Homeland Security Grant Program because the agency does not track the dollar amount spent on medical surge activities. We interviewed officials from ASPR, CDC, and DHS to identify and document criteria and guidance given to state and local entities to plan for medical surge and to learn how federal funds were awarded and utilized.

---

[1] The Joint Commission is an independent, nonprofit organization responsible for evaluating and accrediting over 15,000 U.S. health care organizations and programs, including Department of Defense and Department of Veterans Affairs hospitals.

[2] While the Hospital Preparedness Program awards funds annually to 62 entities—the 50 states, 4 municipalities including the District of Columbia, 5 U.S. territories, and 3 Freely Associated States of the Pacific—we limited our review to the 50 states.

To determine what states have done to prepare for medical surge in a mass casualty event, particularly related to four key components, we obtained and analyzed the 2006 and 2007 ASPR Hospital Preparedness Program cooperative agreement applications and 2006 midyear progress reports (the most current available information—generally effective through March 2007[3]—at the time of our data collection) for the 50 states.[4] We also reviewed the 15 sentinel indicators for the Hospital Preparedness Program.[5] We analyzed the 5 medical-surge-related sentinel indicators for which criteria to evaluate performance were identified and which were reported by the states in their 2006 midyear progress reports. Although ASPR's 2006 guidance for these reports does not provide specific criteria with which to evaluate performance on these indicators, we identified criteria to analyze the data provided for 5 of them from either ASPR's previous program guidance or DHS's *Target Capabilities List*, which includes requirements related to preparing for medical surge.[6] All 5 of the medical-surge-related sentinel indicators we analyzed were related to one of the four key components—hospital capacity. See appendix I for a list of the 15 sentinel indicators. In addition, we obtained and reviewed 20 states' emergency preparedness planning documents relating to medical surge and interviewed state officials from these states regarding their activities related to hospital capacity, alternate care sites, electronic medical volunteer registries, and altered standards of care.[7] We also interviewed these state officials to determine what federal guidance or tools they used and to identify the Department of Defense (DOD) and the Department of Veterans Affairs (VA) hospitals' participation in state planning. Finally, we obtained and reviewed DOD and VA policies and interviewed officials to further understand their policies regarding participation

---

[3] Four of the states we reviewed provided sentinel indicator information as of April 2007, one state as of August 2007, and another state as of September 2007.

[4] The 2006 program year for the Hospital Preparedness Program was September 1, 2006, to August 31, 2007. The 2007 program year is September 1, 2007, to August 8, 2008

[5] Sentinel indicators are smaller component tasks of critical benchmarks, which measure program capacity-building efforts. For example, for the benchmark "Surge Capacity; Beds," one of the sentinel indicators is the number of additional hospital beds for which a recipient could make patient care available within 24 hours. ASPR requires that states report on 15 sentinel indicators, 2 of which are not related to medical surge.

[6] Two of the 15 indicators—total number of hospitals statewide and total population statewide—were used as denominators to analyze the 5 indicators. We were unable to identify or create any usable criteria with which to evaluate the remaining 8 indicators.

[7] We selected the 20 states by identifying 2 states from each of the 10 HHS geographic regions—one with the most ASPR Hospital Preparedness Program funding and one with the least funding. This selection criterion allowed us to take into account population (program funding was awarded using a formula including, in part, population), geographic dispersion, and different geographic risk factors, such as the potential for hurricanes, tornadoes, or earthquakes.

with state and local entities in emergency preparedness planning and responding to mass casualty events.

To determine what concerns states identified as they prepared for medical surge in a mass casualty event, we interviewed emergency preparedness officials from the 20 states and focused our questions on their efforts related to four key components of medical surge we identified. We also asked what further assistance states might need from the federal government to help prepare their health care systems for medical surge.

We did not validate the sentinel indicator data the 50 states reported to ASPR; however, if data for specific indicators were missing or obviously incorrect (e.g., a percentage was greater than 100 percent), we contacted state officials for clarification. We did not examine the accuracy of other self-reported information contained in the midyear progress reports or Hospital Preparedness Program applications from the 20 states we reviewed. During interviews with officials from the 20 states, we discussed the completeness of information provided in their progress reports and applications about four key components related to preparing for medical surge. For each interview, we used a question set that contained open-ended questions. The state emergency preparedness officials we interviewed provided varying levels of detail to answer our questions.

Thus our information from these interviews is illustrative and is intended to provide a general description of what the 20 states have done to prepare for medical surge in a mass casualty event and is not generalizable to all 50 states. We conducted our work from May 2007 through May 2008 in accordance with generally accepted government auditing standards.

APPENDIX III: HOSPITAL PREPAREDNESS PROGRAM FUNDING AND MEDICAL SURGE GUIDANCE AND CONFERENCES

Tables 2, 3, and 4 provide information on ASPR's Hospital Preparedness Program funding and on guidance and other assistance for states to use in preparing for medical surge.

Table 2. ASPR's Hospital Preparedness Program Funding by State, Fiscal Years 2002 through 2007

State	FY 2002	FY 2003	FY 2004	FY 2005	FY 2006	FY 2007[a]
Alabama	$1,972,833	$7,762,315	$7,762,315	$7,326,068	$7,154,927	$6,330,289
Alaska	492,877	1,958,803	1,958,803	1,484,009	1,458,182	1,349,441
Arizona	2,237,637	9,030,450	9,030,450	8,964,023	8,753,827	8,317,173
Arkansas	1,285,691	5,077,591	5,077,591	4,633,962	4,531,309	4,063,403
California	9,962,905	38,773,726	38,773,727	39,203,268	38,325,286	34,106,620
Colorado	1,916,334	7,704,930	7,704,930	7,401,669	7,221,888	6,525,958
Connecticut	1,569,336	6,197,207	6,197,207	5,783,087	5,651,890	4,943,121
Delaware	553,571	2,205,406	2,205,406	1,739,851	1,709,476	1,581,970
Florida	6,441,669	25,775,967	25,775,967	26,311,287	25,638,227	23,432,938
Georgia	3,421,481	13,719,390	13,719,390	13,671,367	13,330,420	12,370,869
Hawaii	719,356	2,856,721	2,856,721	2,407,137	2,345,600	2,129,653
Idaho	751,285	2,998,297	2,998,297	2,572,244	2,521,506	2,359,069
Illinois	3,939,374	15,875,995	15,875,995	15,578,388	14,951,481	13,163,842
Indiana	2,605,616	10,270,929	10,270,929	9,896,622	9,660,723	8,503,785
Iowa	1,383,675	5,436,624	5,436,624	4,965,024	4,846,845	4,280,453

Table 2. Continued

State	FY 2002	FY 2003	FY 2004	FY 2005	FY 2006	FY 2007a
Kansas	1,291,509	5,088,830	5,088,830	4,630,597	4,525,854	4,004,077
Kentucky	1,815,805	7,156,894	7,156,894	6,745,252	6,585,429	5,832,130
Louisiana	1,981,308	7,764,518	7,764,518	7,319,242	7,139,266	5,935,695
Maine	743,913	2,943,648	2,943,648	2,480,391	2,434,432	2,175,388
Maryland	2,301,890	9,150,163	9,150,163	8,855,085	8,645,984	7,619,177
Massachusetts	2,709,678	10,686,180	10,686,180	10,256,868	9,983,770	8,660,567
Michigan	4,100,212	16,141,386	16,141,386	15,787,720	15,395,465	13,298,463
Minnesota	2,155,835	8,542,551	8,542,551	8,173,336	7,983,328	7,050,445
Mississippi	1,352,037	5,327,321	5,327,321	4,869,883	4,759,591	4,189,754
Missouri	2,417,618	9,530,322	9,530,322	9,151,953	8,951,388	7,906,932
Montana	599,516	2,370,015	2,370,015	1,891,709	1,856,928	1,697,530
Nebraska	912,954	3,602,747	3,602,747	3,137,831	3,067,393	2,741,751
Nevada	1,024,136	4,174,253	4,174,253	3,899,038	3,818,014	3,663,636
New Hampshire	728,751	2,905,650	2,905,650	2,452,975	2,404,444	2,166,921
New Jersey	3,509,769	13,878,940	13,878,940	13,601,391	13,269,518	11,560,312
New Mexico	354,709	3,770,553	3,770,553	3,343,195	3,276,757	2,977,887
New York	4,499,138	18,019,873	18,019,873	17,757,875	16,937,704	14,561,258
North Carolina	3,368,351	13,417,400	13,417,400	13,251,044	12,948,887	11,727,581
North Dakota	498,792	1,963,221	1,963,221	1,461,290	1,435,800	1,306,102
Ohio	4,648,274	18,234,914	18,234,914	17,843,984	17,397,207	15,050,914
Oklahoma	1,586,804	6,250,131	6,250,131	5,825,603	5,681,308	5,037,444
Oregon	1,575,470	6,255,978	6,255,978	5,898,716	5,767,951	5,191,530
Pennsylvania	5,007,754	19,616,940	19,616,940	19,254,011	18,776,677	16,271,242
Rhode Island	656,125	2,603,466	2,603,466	2,132,147	2,089,651	1,853,432
South Carolina	1,804,277	7,146,769	7,146,769	6,789,755	6,632,258	5,978,140
South Dakota	542,431	2,147,489	2,147,489	1,659,192	1,630,322	1,491,255
Tennessee	2,454,062	9,699,934	9,699,934	9,359,882	9,138,647	8,155,520
Texas	8,328,119	33,338,368	33,338,368	34,045,388	33,177,278	30,301,320
Utah	1,115,143	4,448,125	4,448,125	4,066,334	3,978,558	3,732,769
Vermont	485,864	1,927,552	1,927,552	1,438,965	1,415,048	1,290,942
Virginia	2,992,259	11,890,053	11,890,053	11,701,905	11,387,068	10,189,048
Washington	2,533,418	10,069,141	10,069,141	9,799,166	9,562,647	8,608,090
West Virginia	950,564	3,725,218	3,725,218	3,245,672	3,176,132	2,805,313
Wisconsin	2,327,920	9,180,227	9,180,227	8,799,529	8,588,953	7,544,102
Wyoming	441,296	1,747,144	1,747,144	1,260,221	1,241,982	1,152,882
Total	$113,669,341	$450,360,266	$450,360,266	$434,115,151	$423,163,319	$377,188,133

Source: ASPR.

[a] The fiscal year 2007 funds for the Hospital Preparedness Program were awarded to the states in September 2007. States can expend these funds during the 2007 program year, which runs from September 1, 2007, to August 8, 2008.

Table 3. Federal Guidance and Technical Assistance Published for States to Use in Preparing for Medical Surge

Federal guidance

National Response Framework

Target Capabilities List

Homeland Security Presidential Directive 10 (HSPD-10) Biodefense for the 21st Century

Homeland Security Presidential Directive 21 (HSPD-21) Public Health and Medical Preparedness

HHS planning guidance

State & Local Pandemic Influenza Planning Checklist

Law Enforcement Pandemic Influenza Planning Checklist

Correctional Facilities Pandemic Influenza Planning Checklist

Draft Guidance on Allocating and Targeting Pandemic Influenza Vaccine

Interim Public Health Guidance for the Use of Facemasks and Respirators in Non-Occupational Community Settings during anInfluenza Pandemic (CDC)

www.Pandemicflu.gov

In a Moment's Notice: Surge Capacity for Terrorist Bombings (CDC)

ASPR programs

Hospital Preparedness Program

Healthcare Facilities Partnership Program

Healthcare Facilities Emergency Care Partnership Program

Emergency System for Advance Registration of Volunteer Health Professionals (ESAR-VHP) Program

Bioterrorism Training Curriculum Development Program

Regional Emergency Coordinators

ASPR planning documents and handbooks

HHS Plan to Combat Bioterrorism and Other Public Health Emergencies

Medical Surge Capacity and Capability Handbook

Emergency Management Assistance Compact Quick Tips

Interim Public Health and Healthcare Supplement to the National Preparedness Goal

AHRQ tools and resources

Adapting Community Call Centers for Crisis Support[a]

Altered Standards of Care in Mass Casualty Events[a]

Bioterrorism and Emerging Infections Site

Table 3. Continued

CBRNE (Chemical, Biological, Radiological, Nuclear, and Explosive) Events (Questionnaire)
Community-Based Mass Prophylaxis: A Planning Guide for Public Health Preparedness
Computer Staffing Model for Bioterrorism Response—BERM Version 2
Decontamination of Children: Preparedness and Response for Hospital Emergency Departments: Video
Emergency Preparedness Atlas: U.S. Nursing Home and Hospital Facilities[a]
Emergency Preparedness Resource Inventory (EPRI)
Emergency Severity Index
Evaluation of Hospital Disaster Drills: A Module-Based Approach
Health Emergency Assistance Line and Triage Hub (HEALTH) Model
Nursing Homes in Public Health Emergencies[a]
Pediatric Terrorism and Disaster Preparedness: A Resource for Pediatricians
Personal Protective Equipment, Decontamination, Isolation/Quarantine, and Laboratory Capacity
Predicting Health Care Use Resulting From Terrorism: Tools To Aid State Planning: Summary
Mass Medical Care with Scarce Resources: A Community Planning Guide[a]
National Hospital Available Beds for Emergencies and Disasters (HAvBED) System: Final Report and Appendixes[a]
Project XTREME: Cross-Training Respiratory Extenders for Medical Emergencies[a]
Readiness and Response to Public Health Emergencies: Help Needed Now From Professional Nursing Associations
Reopening Shuttered Hospitals to Expand Surge Capacity (Surge Toolkit and Facility Checklist)[a]
Rocky Mountain Regional Care Model for Bioterrorist Events (Alternate Care Site Selection Tool)[a]
Standardized Hospital Bed Definitions (HAvBED)
Understanding Needs for Health System Preparedness and Capacity for Bioterrorist Attacks (Questionnaire)
AHRQ issue briefs
Addressing the Smallpox Threat: Issues, Strategies, and Tools
Disaster Planning Drills and Readiness Assessment
Optimizing Surge Capacity: Hospital Assessment and Planning
Optimizing Surge Capacity: Regional Efforts in Bioterrorism Readiness
The Role of Information Technology and Surveillance Systems in Bioterrorism Readiness

Table 3. Continued

Bioterrorism and Other Public Health Emergencies: Linkages with Community Providers
Surge Capacity—Education and Training for a Qualified Workforce
Surge Capacity: Facilities and Equipment
Addressing Surge Capacity in a Mass Casualty Event
Mass Prophylaxis: Building Blocks for Community Preparedness
Developing Alternative Approaches to Mass Casualty Care

AHRQ evidence reports

Evidence Report/Technology Assessment: Number 51: Training of Clinicians for Public Health Events Relevant to Bioterrorism Preparedness
Evidence Report/Technology Assessment: Number 59: Bioterrorism Preparedness and Response: Use of Information Technologies and Decision Support Systems
Evidence Report/Technology Assessment: Number 95: Training of Hospital Staff To Respond to a Mass Casualty Incident
Evidence Report/Technology Assessment: Number 96: Regionalization of Bioterrorism Preparedness and Response
Evidence Report/Technology Assessment: Number 141: Pediatric Anthrax: Implications for Bioterrorism Preparedness

RAND reports and tools

Improving and Enhancing Telephone-based Disease Surveillance Systems in Local Health Departments[b]
Assessing Public Health Emergency Preparedness: Concepts, Tools, and Challenges
Enhancing Public Health Preparedness: Exercises, Exemplary Practices, and Lessons Learned, Phase III[b]
Quality Improvement - Implications for Public Health Preparedness[b]
Public Health Preparedness - Integrating Public Health and Hospital Preparedness[b]
Organizing State and Local Health Departments for Public Health Preparedness[b]
Tabletop Exercises for Pandemic Influenza Preparedness in Local Public Health Agencies[b]
Facilitated Look Backs - A New Quality Improvement Tool for Management of Routine Annual and Pandemic Influenza[b]
Enhancing Public Health Preparedness: Exercises, Exemplary Practices, and Lessons Learned[b]
Exemplary Practices in Public Health Preparedness[b]

Table 3. Continued

Learning From Experience: The Public Health Response to West Nile Virus, SARS, Monkeypox, and Hepatitis A Outbreaks in the United States[b]
Test to Evaluate Public Health Disease Reporting Systems in Local Public Health Agencies[b]
Bioterrorism Preparedness Training and Assessment Exercises for Local Public Health Agencies[b]
Tabletop Exercise for Pandemic Influenza Preparedness in Local Public Health Agencies[b]
Quality Improvement: Implications for Public Health Preparedness[b]
Designing and Conducting Tabletop Exercises to Assess Public Health Preparedness for Manmade and Naturally Occurring Biological Threats
National Highway Traffic Safety Administration resources
EMS Pandemic Influenza Guidelines for Statewide Adoption
Recommendations for Protocol Development for 9-1-1 Personnel

Source: HHS.
[a] Projects for which ASPR awarded funding and provided support and direction.
[b] HHS contracted with RAND to produce these reports and tools.

Table 4. Federal Conferences and Meetings with States that Provided Information to Prepare for Medical Surge

HHS conference
Pandemic Planning: A Convening of the States, December 5, 2005
ASPR conferences
Annual Hospital Preparedness Program Cooperative Agreement Recipient Conference
Annual ESAR-VHP Conference
2008 Bioterrorism Training and Curriculum Development Program Conference
AHRQ conferences
Public Health Emergencies: Strategies and Tools for Meeting the Needs of Children, Web Conference, January 11, 2006
Mass Casualty Care: Overlooked Community Resources, Web Conference, May 17, 2005
Addressing Surge Capacity in a Mass Casualty Event, Web Conference, October 26, 2004
Surge Capacity and Health System Preparedness: Facilities and Equipment, Web Conference, July 13, 2004
Surge Capacity and Health System Preparedness: Education and Training for a Qualified Workforce, Web Conference, March 2, 2004

Table 4. Continued

Bioterrorism and Other Public Health Emergencies—Linkages with Community Providers, Web Conference, December 16, 2003

The Role of Information/Communication Technology and Monitoring/Surveillance Systems in Bioterrorism Preparedness, Web Conference, October 21, 2003

Surge Capacity Assessments and Regionalization Issues, Web Conference, June 17, 2003

Disaster Planning Drills and Readiness Assessment, Web Conference, April 15, 2003

Addressing the Smallpox Threat: Issues, Strategies, and Tools, Web Conference, March 3, 2003

Bioterrorism and Health System Preparedness, Web Conference, April 29, 30, and May 1, 2002

Expert Meeting on Bioterrorism, February 18, 2000

Source: HHS.

APPENDIX IV: DATA FOR THE FIVE SURGE-RELATED SENTINEL INDICATORS FOR HOSPITAL CAPACITY FROM THE HOSPITAL PREPAREDNESS PROGRAM

Figures 1 through 5 provide data for the five surge-related sentinel indicators for hospital capacity from ASPR's Hospital Preparedness Program 2006 midyear progress reports.

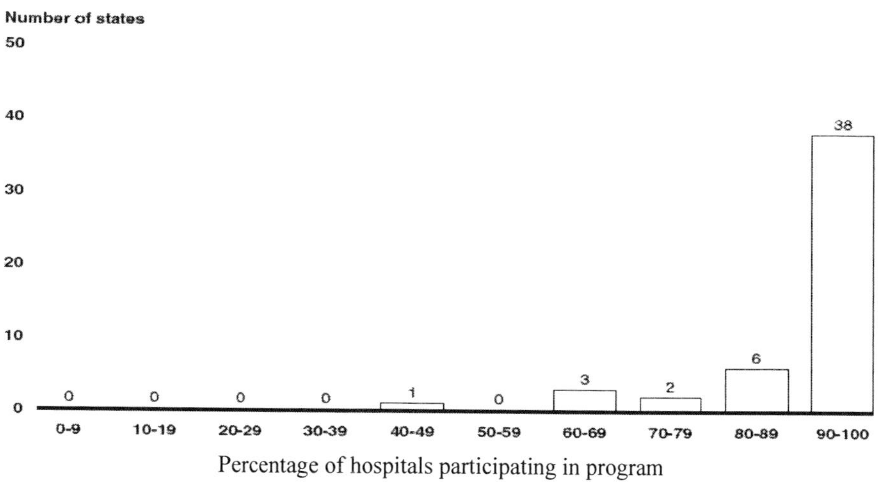

Source: GAO analysis of ASPR data.

Figure 1. Hospital Participation in Individual States' Hospital Preparedness Programs.

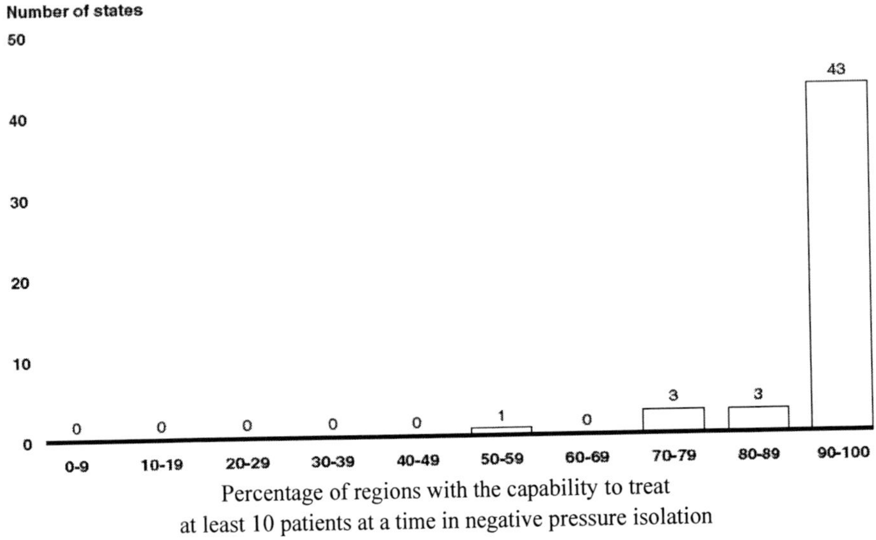

Source: GAO analysis of ASPR data.

Figure 2. States Whose Regions Have the Capability to Treat at Least 10 Patients at a Time in Negative Pressure Isolation.

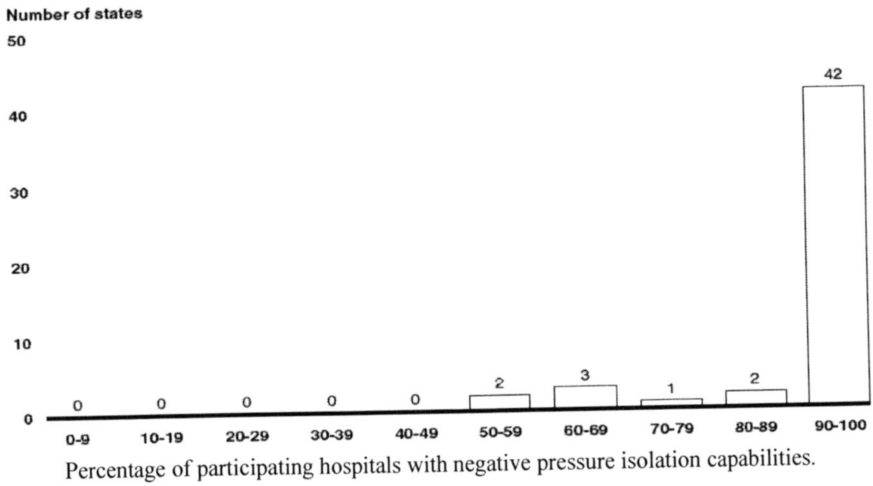

Source: GAO analysis of ASPR data.

Figure 3. States with Participating Hospitals That Have Negative Pressure Isolation Capabilities.

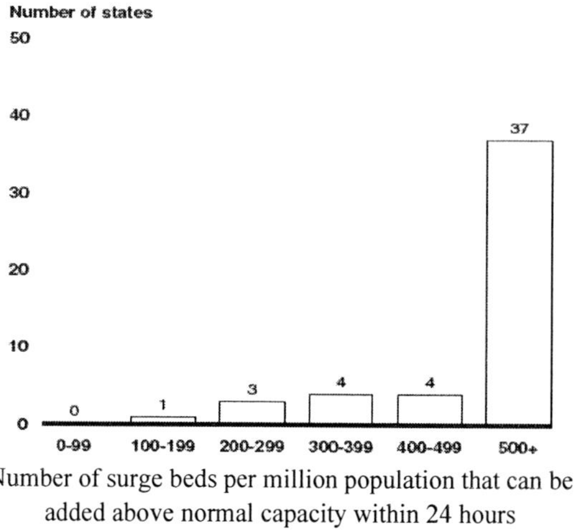

Source: GAO analysis of ASPR data.

Note: One state reported that it did not know the number of surge beds that could be added.

Figure 4. Number of Additional Surge Beds per Million Population That Can Be Added above Normal Capacity within 24 Hours.

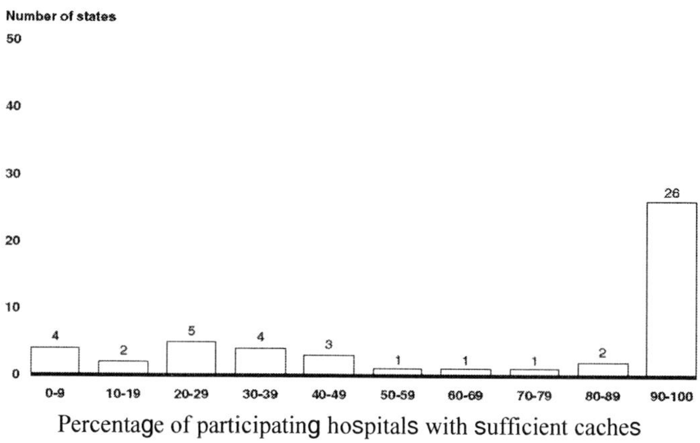

Source: GAO analysis of ASPR data.

Note: One state reported that it did not know the number of sufficient pharmaceutical caches that were available because it was unsure of the definition of "sufficient."

Figure 5. States Whose Participating Hospitals Have Sufficient Pharmaceuticals to Treat Hospital Personnel and Their Family Members.

APPENDIX V: COMMENTS FROM THE DEPARTMENT OF HEALTH AND HUMAN SERVICES

THE SECRETARY OF HEALTH AND HUMAN SERVICES
WASHINGTON, DC 20201

MAY 23 2008

Cynthia A. Bascetta
Director, Health Care
U.S. Government Accountability Office
Washington, DC 20548

Dear Ms. Bascetta:

Enclosed are the Department's comments on the U.S. Government Accountability Office's (GAO) draft report entitled, "Emergency Preparedness: States are Planning for Medical Surge, but Could Benefit from Shared Guidance for Allocating Scarce Medical Resources" (GAO-08-668).

The Department appreciates the opportunity to review and comment on this draft before its publication.

Sincerely,

Jennifer R. Luong
for Vincent J. Ventimiglia, Jr.
Assistant Secretary for Legislation

Attachment

COMMENTS OF THE DEPARTMENT OF HEALTH AND HUMAN SERVICES (HHS) ON THE GOVERNMENT ACCOUNTABILITY OFFICE'S (GAO) DRAFT REPORT ENTITLED: "EMERGENCY PREPAREDNESS: STATES ARE PLANNING FOR MEDICAL SURGE, BUT COULD BENEFIT FROM SHARED GUIDANCE FOR ALLOCATING SCARCE MEDICAL RESOURCES" (GAO 08-668)

Overall, the report is a fair representation of progress that has been made to improve medical surge capacity since 2001. This report focuses on the activities and concerns of State agencies and officials. Since "all disasters are local", it might be more worthwhile to have some local perspectives included as well.

The program name for ESAR-VHP is incorrect throughout the report. The program name is the "Emergency System for Advance Registration of Volunteer Health Professionals". "Systems" should be changed to "System" in each occurrence.

APPENDIX VI: COMMENTS FROM THE DEPARTMENT OF HOMELAND SECURITY

U.S. Department of Homeland Security
Washington, DC 20528

Homeland Security

May 16, 2008

Ms. Cynthia Bascetta
Director, Health Care
U.S. Government Accountability Office
441 G St. NW
Washington, D.C. 20548

Dear Ms. Bascetta:

Thank you for the opportunity to review and provide comments on the Government Accountability Office's (GAO) draft report entitled, *Emergency Preparedness: States Are Planning for Medical Surge, but Could Benefit from Shared Guidance for Allocating Scarce Medical Resources* (GAO-08-668).

The Department of Homeland Security (DHS) has reviewed the referenced GAO Report, and we concur with the findings. We would like to offer a comment on the issue of altered standards of care. GAO offers a definition of "standard of care" in footnote number 4 (pg. 2) which we feel to be accurate and generally accepted. Our concern is that the discussion as written is not one of altered standards of care, but rather the public's need to accept an altered expectation of care. If, as defined in footnote number 4, a standard of care is "how similarly qualified health care providers would manage the patient's care under the same or similar circumstances" then the "standard" of care does not change, but rather it is the type, or level, of care which is altered. While this might be construed as semantics, it highlights the need for enhanced pre-event public messaging to prepare the population for a different look to health care in a mass casualty incident.

Another point of concern is in the recommendation that the Department of Health and Human Services (HHS) provide best practices, or passive guidance to states when considering rationing of scarce resources. Upon review, it seems that states are having difficulty in getting past the ethical and legal issues surrounding rationing. Further exploration may be needed regarding the specific facets of the issue and the possible production of guidance to direct states' discussion on rationing of scarce resources.

DHS is dedicated to assisting our state partners in maintaining the health and resiliency of the homeland. Thank you for the opportunity to review and provide comments on this draft report.

Sincerely,

Penelope G. McCormack
Acting Director
Departmental GAO/OIG Liaison Office

APPENDIX VII: COMMENTS FROM THE DEPARTMENT OF DEFENSE

THE ASSISTANT SECRETARY OF DEFENSE

1200 DEFENSE PENTAGON
WASHINGTON, DC 20301-1200

HEALTH AFFAIRS

MAY 2 1 2008

Ms. Cynthia A. Bascetta
Director, Health Care
U.S. Government Accountability Office
441 G Street, N.W.
Washington, DC 20548

Dear Ms. Bascetta,

This is the Department of Defense response to the Government Accountability Office (GAO) draft report, GAO-08-668, "EMERGENCY PREPAREDNESS: States Are Planning for Medical Surge, but Could Benefit from Shared Guidance for Allocating Scarce Medical Resources," dated May 1, 2008 (GAO Code 290623).

Thank you for the opportunity to comment on this important report. I concur with the GAO's conclusions and findings. The draft report reflects the issues and narrative that my staff exchanged with the audit team during the research phase of the audit.

My points of contact on this matter are Captain D. W. Chen (Functional), who can be reached at (703) 845-3376, and Mr. Gunther Zimmerman (Audit Liaison), who can be reached at (703) 681-4360.

Sincerely,

for S. Ward Casscells, MD

APPENDIX VIII: COMMENTS FROM THE DEPARTMENT OF VETERANS AFFAIRS

THE DEPUTY SECRETARY OF VETERANS AFFAIRS
WASHINGTON

May 19, 2008

Ms. Cynthia A. Bascetta
Director
Health Care
U. S. Government Accountability Office
441 G Street, NW
Washington, DC 20548

Dear Ms. Bascetta:

The Department of Veterans Affairs (VA) has reviewed your draft report, ***EMERGENCY PREPAREDNESS: States Are Planning for Medical Surge, but Could Benefit from Shared Guidance for Allocating Scarce Medical Resources*** (GAO-08-668) and agrees with your findings as they pertain to VA.

The Veterans Health Administration (VHA) and its medical centers commit to providing emergency aid and supplies during mass casualty events to the best of their ability. VA medical centers are required to meet their primary, Congressionally-mandated, mission of delivery of health care to veterans and others during times of emergencies. VA medical centers would assess Emergency Support Function (ESF) #8 requests for public health and medical emergency aid in the context of that which is needed to meet VA's primary functions. In addition to that mandate, VA cannot commit in advance to providing emergency aid and supplies due to a number of factors:

1. Not all VA medical centers provide emergency services nor do they have the same level of emergency supplies.

2. VA medical centers vary widely in size and complexities. Accordingly, VA's capability to provide emergency care and support varies widely.

3. Requiring all VA medical centers to maintain the same level of emergency care and capabilities is impractical and cost prohibitive.

In addition, all VA medical centers are accredited by The Joint Commission (TJC), which includes compliance with Emergency Management (EM) standards under the Environment of Care (EOC) umbrella. TJC evaluates each facility no less than every 3 years, and EM compliance is included. This evaluation is based in part on the VA medical center's capability and expectations to provide EM services within its community. GAO's findings on pages 19-20, which indicate inconsistencies from State to State regarding VA medical centers' stance towards treating non-veterans in the event of a medical surge, likely stem from VA medical centers' varying capabilities to provide emergency medical treatment in any given State.

Ms. Cynthia A. Bascetta

The GAO draft report (pages 19-20) indicates that a number of States reported that VA facilities were either unwilling to treat nonveteran patients or lacked the authority to do so. Footnote 95 addresses this misconception by stating that VA is authorized to provide emergency care to non-veterans on a humanitarian basis. In addition, however, VA has specific statutory authority to provide care and services during certain disasters and emergencies. That authority is codified in section 1785 of title 38, United States Code. Regulations implementing this authority were published in the Federal Register on May 12, 2008, (73 Fed. Reg. 26945) and will be codified at 38 C.F.R. § 17.86.

In addition to specific authorities in title 38, United States Code, VA also responds to disasters and emergencies pursuant to the authority in the Robert T. Stafford Disaster Relief and Emergency Assistance Act, 42 U.S.C. §§ 5121, *et seq.* (the Stafford Act). The Stafford Act sets forth the Federal government's authority to respond to major disasters and emergencies.

The National Response Framework (NRF) is the Federal-wide document describing how the Federal government will assist State and local responses in major disasters and emergencies. It provides a process and a structure for the systematic and coordinated Federal response. The Plan organizes the types of Federal response assistance under Emergency Support Functions (ESF). Each ESF groups together, by function, the types of direct Federal Assistance that a State is most likely to need (*e.g.*, mass care, health and medical services) as well as the kinds of support that are necessary to sustain the Federal response (*e.g.*, transportation, communications). For each of these support functions, the NRF designates a "primary agency" which is responsible for planning and coordinating the delivery of the function. In addition, the NRF designates "support agencies" for each function. VA has a specific support role in ESF #3 (Public Works and Engineering), #5 (Emergency Management), #6 (Mass Care, Emergency Assistance, Housing, and Human Services), #7 (Logistics Management and Resource Support) and #8 (Public Health and Medical Services). Under the auspices of the NRF, VA may be directed to provide assistance to State and local relief efforts.

VA appreciates the opportunity to comment on your draft report.

Sincerely yours,

Gordon H. Mansfield

ENDNOTES

[1] By health care systems, we mean both public health and medical systems, including hospitals.

[2]A bioterrorism attack is the deliberate release of viruses, bacteria, or other germs (agents) used to cause illness or death in people, animals, or plants. These agents are typically found in nature, but it is possible that they could be changed to increase their ability to cause disease, to make them resistant to current medicines, or to increase their ability to be spread into the environment. Biological agents can be spread through the air, through water, or in food.

[3]A mass casualty event is a public health or medical emergency that could involve thousands, or even tens of thousands, of injured or ill victims.

[4]A standard of care is the diagnostic and treatment process that a provider should follow for a certain type of patient or illness, or certain clinical circumstances. It is how similarly qualified health care providers would manage the patient's care under the same or similar circumstances.

[5] Alternate care sites deliver medical care outside of hospital settings for patients who would normally be treated as inpatients.

[6] The term "altered standards" generally means a shift to providing care and allocating scarce equipment, supplies, and personnel in a way that saves the largest number of lives, in contrast to the traditional focus of treating the sickest or most injured patients first. For example, it could mean applying principles of field triage to determine who gets what kind of care, changing infection control standards to permit group isolation rather than single- person isolation units, changing who provides various kinds of care, or changing privacy and confidentiality protections temporarily.

[7] See Pub. L. No. 107-296, 116 Stat. 2135 (2002).

[8] Pub. L. No. 109-417, § 101, 120 Stat. 2831, 2832 (2006) (codified at 42 U.S.C. § 300hh).

[9] The National Response Framework details the missions, policies, structures, and responsibilities of federal agencies for coordinating resource and programmatic support to states, tribes, and other federal agencies.

[10] DOD Directive 3025.1, §§ 4.6.1.2. and 4.5.1 *Military Support to Civil Authorities* (Jan. 15, 1993).

[11] The Joint Commission is an independent, nonprofit organization that evaluates and accredits more than 15,000 U.S. health care organizations and programs, including DOD and VA hospitals.

[12] The 2006 program year for the Hospital Preparedness Program was September 1, 2006, to August 31, 2007. The 2007 program year is September 1, 2007, to August 8, 2008.

[13] While the Hospital Preparedness Program awards funds annually to 62 entities—the 50 states; 4 municipalities, including the District of Columbia; 5 U.S. territories; and 3 Freely Associated States of the Pacific—we limited our review to the 50 states.

[14] Sentinel indicators are smaller component tasks of critical benchmarks, which measure program capacity-building efforts such as purchasing equipment and supplies and acquiring personnel. For example, for the benchmark "Surge Capacity; Beds," one of the sentinel indicators is the number of additional hospital beds for which a recipient could make patient care available within 24 hours. ASPR requires that states report on 15 sentinel indicators.

[15] Two of the 15 indicators—total number of hospitals statewide and total population statewide—were used as denominators to analyze the 5 indicators.

[16] Before 2008, the Homeland Security Grant Program consisted of five programs, including the four mentioned and the Law Enforcement Terrorism Prevention Program. In 2008 the Law Enforcement Terrorism Prevention Program was rolled into the State Homeland Security Program and the Urban Areas Security Initiative.

[17] Prior to March 2007, the Hospital Preparedness Program was administered by HHS's Health Resources and Services Administration (HRSA) and was named the National Bioterrorism Hospital Preparedness Program.

[18] The four municipalities are the District of Columbia, New York City, Chicago, and Los Angeles County; the five U.S. territories are Puerto Rico, American Samoa, Guam, the Northern Marianas Islands, and the U.S. Virgin Islands; the three Freely Associated States of the Pacific are the Federated States of Micronesia, Marshall Islands, and Palau.

[19] Negative pressure isolation rooms maintain a flow of air into the room to ensure that contaminants and pathogens cannot escape from the room to other parts of the facility and to protect the health of workers and other patients.

[20] A medical volunteer is a professional who renders aid or performs health services voluntarily, without pay or remuneration

[21] Credentialing is the process of obtaining, verifying, and assessing the qualifications of a health care professional. In an emergency, a single set of standards and definitions can be assigned to medical volunteers so the volunteers can render services across communities and state lines.

[22] The Joint Commission has standards that require hospitals to establish organization- specific staff-to-patient ratios based upon the organization's assessment of patient care needs. The assessment usually involves consideration of numbers, types, and seriousness of illness of various patient groups.

[23] An additional $218 million was provided to four large municipalities, five U.S. territories, and three Freely Associated States of the Pacific for a total of approximately $2.5 billion. Over the 2-year period, fiscal years 2004 and 2005, HHS also awarded an additional $200,000 to 48 states for ESAR-VHP development through this program.

[24] Since January 2006, HHS also had awarded the 62 recipients an additional $400 million in two phases and a supplement to prepare for a pandemic influenza outbreak. The funds were awarded to accelerate their current planning efforts for an influenza pandemic and to exercise their plans. These funds included $75 million in August 2007 that could be used, in part, to develop pandemic alternate care sites and to conduct medical surge exercises.

[25] For example, one of the activities is to receive and treat surge casualties. One of the critical tasks associated with this activity is to ensure adequacy of medical equipment and supplies in support of immediate medical response operations and for restocking requested supplies and equipment.

[26] ESAR-VHP guidelines assign each volunteer to one of four emergency credentialing levels depending on the medical credentials possessed and verified. For example, Level 4 is assigned to volunteers who have been registered into the system, without having any credentials, such as licenses, certifications, and hospital privileges, verified. Level 1 is assigned to volunteers whose credentials have been fully verified.

[27] The number of available beds refers to the number of beds that are licensed, physically available, and have staff on hand to attend to the patient who occupies the bed. These beds must include supporting space, equipment,

medical supplies, ancillary and support services, and staff to operate under normal circumstances.

[28] The RAND Corporation is a nonprofit institution that conducts research and issues reports on social and economic issues, such as education, poverty, crime, and the environment, as well as a range of national security issues, including emergency preparedness.

[29] The 2006 program year was from September 1, 2006, to August 31, 2007; therefore, information provided in the midyear progress reports was reported as of March 2007.

[30] Four of the states we reviewed provided sentinel indicator information as of April 2007, one state as of August 2007, and another state as of September 2007.

[31] Among other standards, HAvBED systems are required to report on seven categories of staffed available beds. The seven bed categories are intensive care, medical and surgical, burn, pediatric intensive care, pediatric, psychiatric, and negative pressure isolation. HAvBED systems are also required to report on emergency department diversions, decontamination facilities available, and ventilators available. ASPR allows each state to use Hospital Preparedness Program funds to develop its own bed tracking system as long as the system meets HAvBED requirements.

[32] In addition to hospitals, these facilities could include skilled nursing facilities, assisted living facilities, and residential treatment facilities.

[33] ASPR requires all recipients to complete the development of their bed tracking system by August 8, 2008.

[34] Directive 3025.1, Section 4.5.1 authorizes military officials to take necessary actions to respond to civilian requests for assistance in emergencies, which may include accepting civilian patients. This decision can be authorized by DOD or, in cases of urgent need, by the commander of the local military hospital.

[35] VA is authorized to furnish hospital care or medical services as a humanitarian service to non-VA beneficiaries in emergency cases. See 38 U.S.C. § 1784; 38 C.F.R. §§ 17.37, 17.43, 17.95, 17. 102. VA is also authorized to provide care and services during certain disasters and emergencies. See 38 U.S.C. § 1785; 38 C.F.R. § 17.86.

[36] According to a VA General Counsel memorandum (Guidance on Entering into Mutual Aid Agreements, July 23, 2003), hospitals can also enter into mutual aid agreements in which VA hospitals and local entities agree to assist each other during disasters and emergencies. These agreements often include provisions to accept patients from other hospitals if the transferring hospital

has an overwhelming number of patients or if the transferring facility does not have the resources for patients who require specialized medical treatment. However, these mutual aid agreements must state that the agreement is limited by certain VA obligations that may take precedence over the agreement to assist local hospitals during an emergency, such as its obligations under the National Disaster Medical System and its obligations to assist DOD during a time of war or national emergency.

[37] The Modular Emergency Medical System provides detailed standards for a system of medical care that can be expanded as the need arises. It provides a framework for the organization of care, particularly for setting up predetermined, special-use alternate care sites. It provides information on what general kinds of care are provided and where and who will provide care.

[38] The seven categories of health care professionals are physicians, registered nurses, marriage and family therapists, medical and public health social workers, mental health and substance abuse social workers, psychologists, and mental health counselors.

[39] A ventilator mechanically moves oxygen into and out of the lungs of a patient who is physically unable to breathe on his or her own, or whose breathing is insufficient to maintain life.

[40] The pandemic influenza Web site can be accessed at www.pandemicflu.gov. This Web site is managed by HHS.

[41] Officials from the remaining state reported that they did not know how many beds were available statewide above the current daily staffed bed capacity.

[42] Centers for Medicare & Medicaid Services (CMS), through the Medicare and Medicaid programs, is a significant source of reimbursement for medical services, including those provided in hospital settings. Additionally, private insurers typically are guided by CMS policies regarding reimbursement.

[43] On CMS's Web site, the agency provides some broad guidance on its role during an emergency. See http://www.cms.hhs.gov/SurveyCertEmergPrep, *Provider Survey and Certification Frequently Asked Questions, Declared Public Health Emergencies - All Hazards Health Standards and Quality Issues* (Sept. 30, 2007).

[44] Pub. L. No. 104-191, 110 Stat. 1936; 45 C.F.R. Parts 160 and 164.

[45] Pub. L. No. 99-272, 100 Stat. 164 (1986) (codified, as amended, at 42 U.S.C. § 1395dd).

[46] An individual's authorization is not required to use and disclose protected health information for some purposes, such as treatment, payment, and health care operations. 45 C.F.R. § 164.506(c).

[47] Social Security Act § 1135(b) (codified at 42 U.S.C. §1320b-5). These waivers are limited to a 72-hour period that begins when a hospital implements a disaster protocol.

[48] See http://www.hhs.gov/katrina/ssawaiver.html.

[49] MRC is a federal program within the U.S. Surgeon General's Office, which is in HHS. MRC units are community-based and organize and utilize volunteers to, among other things, prepare for and respond to emergencies. MRC volunteers include medical and public health professionals as well as other community members such as interpreters and legal advisers.

[50] DMAT is an HHS program in which volunteer medical personnel provide medical care during a disaster. DMATs supplement local medical care until other federal resources can be mobilized and deployed to disaster sites.

[51] By comparison, seasonal influenza in the United States generally results in 200,000 hospitalizations annually.

[52] Department of Health and Human Services, *HHS Pandemic Influenza Plan* (Washington, D.C.: Nov. 2005).

[53] The group brought together experts in law, medicine, policy making, and ethics with representatives from medical facilities and city, county, and state government.

[54] The task force included officials from DHS, HHS, ASPR, CDC, DOD, and VA. See Asha V. Devereaux, et al., "Definitive Care for the Critically Ill During a Disaster: A Framework for Allocation of Scarce Resources in Mass Critical Care: From a Task Force for Mass Critical Care Summit Meeting, January 26 to 27, 2007, Chicago, Il.," *Chest* (2008): 133, 51-66.

INDEX

H

I

J

K

L

Q

R

S

T

U

V

W